Real Science, Real Decisions

A Collection of Thinking Activities from *The Science Teacher*

National Science Teachers Association

Second Printing 1993
by the **National Science Teachers Association**
1840 Wilson Blvd.
Arlington, VA 22201-3000

Stock Number PB082X
ISBN 0-87355-097-8
Printed in the United States of America

Real Science, Real Decisions

A Collection of Thinking Activities from *The Science Teacher*

The Science Teacher **is published monthly, September through May. For subscription information, write to Member Services, National Science Teachers Association, 1840 Wilson Blvd., Arlington, VA 22201-3000.**

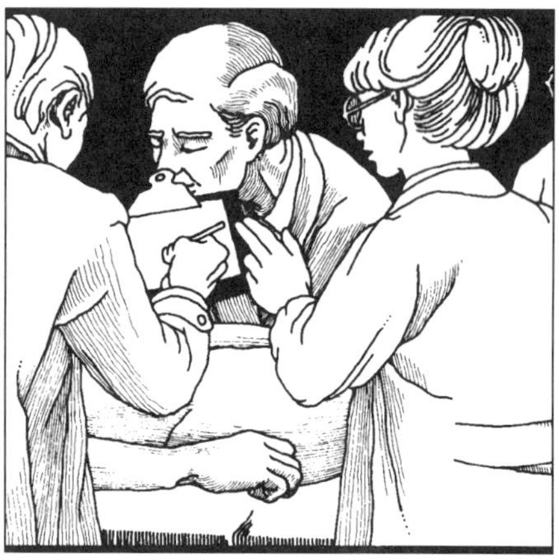

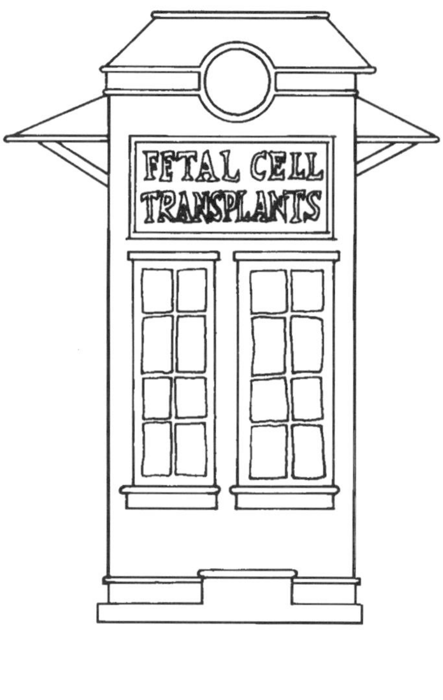

Introduction

There is hardly a unit we teach anymore in science that does not collide with life outside the laboratory and classroom. Modern biology seems to be boiling with issues that confront the ethical and moral values of our fast-changing society. The physical sciences are generating technology that transforms our world faster than most college-educated citizens can comprehend. The decisions our students must make about what to do with the consequences of new scientific discoveries are difficult and seemingly endless: Should Americans spend trillions of tax dollars on a super collider that promises only new knowledge? Are the sports of hunting and fishing moral behavior for modern humans? How must human fetal cell transplant research be regulated?

The purpose of this collection of Difficult Decisions is for teachers to: recognize that useful social situations arise continually from scientific research; capitalize on the personal, social, and technological interests of science students to address these issues; and practice teaching strategies that maximize the thoroughness with which students grapple with the opposing positions of critical issues.

Each difficult decision is a separate lesson providing materials for teachers and students. The teacher's section is an extensive presentation of each issue: its scientific and technological roots, and its personal and social impact. Higher level questions enable teachers to extend the activity beyond a single class period. A photocopyable students' page follows and presents the issue with a question, illustration, and scenario. Additional questions probe various parts of the issue. Although references were current at the time of publication, the burning issues in science change daily and many of today's questions may soon disappear. But new issues about science and society will take their place and burn with relevance. Thus the difficult decisions appearing in this soon-to-be-dated collection should be used as instructional models.

Strategies for engaging science students in the analysis of issues are not well established in science teaching. Several effective techniques for involving students have individuals or groups engage one another on opposing sides of propositions. Debates or panel discussions enable students to prepare formal arguments. Alternatively, teachers may group students into several committees, each of which shares its knowledge and ideas about the resolution of the four or five questions asked on the student page of each issue. For issues carried beyond the awareness level of a single lesson, a teacher may bring in local speakers as a way to engage students with the community, or fashion surveys to compare class and community positions on an issue, or view a professional video to measure if the video has an effect in changing attitudes.

At their best, these chapters will awaken some students to the real decisions that result from real science.

Difficult Decisions: Fetal Cell Transplants

by Irwin L. Slesnick and Jal S. Parakh

Transplanting embryonic cells from fetuses into adults as a means of rejuvenating diseased organs poses profound ethical and moral challenges. Recently, fetal nerve and glandular tissues have been transplanted into the brain and pancreas of Parkinson's and diabetes patients, respectively, with mixed results. The fetal cells used in transplants are supplied by the 1.5 million fetuses aborted each year in the United States. Experimentation with human fetal tissue raises such issues as the right of a woman to decide how the cells of her aborted fetus are used, the involvement of physicians and clinics in marketing fetal cells and transplant techniques, and the morality of using cells acquired through induced abortion.

Background

Cells in the body of developing young mammals gradually change from a stage of nondifferentiation to a stage of irreversible commitment to a specific form and function. For example, undifferentiated cells from the brain of an early fetus do not yet "know" that they are brain cells. Transplanted into the brain of an adult, they produce tumor-like growths. Brain cells from fetuses older than three months are differentiated and have already started to develop axons and dendrites. By now they have lost the capacity to regenerate parts. Brain cells of 11-week-old fetuses, however, do "know" they are brain cells but have not yet differentiated axon and dendrite processes and the synaptic connections of functioning cells. Therefore, when transplanted into adult brains, these cells can integrate and mature with the tissue in the brain of an adult host. They stimulate the growth of such supporting tissues as blood vessels. As they mature further they can secrete the hormones essential to normal functioning.

—Art by Max-Karl Winkler

Another characteristic of fetal cells transplanted into adults is their general acceptance by the immune system of the host. When the donor fetal cells are from a species different than the host, however, rejection usually occurs unless suppressor drugs are used.

While aborted fetuses are the current source of embryonic cells for research in cell division, differentiation, and transplantation; the future source for cells will likely be the cultured tissues of a few fetuses frozen for convenient storage and ease of transportation.

Medical technology

Techniques for transplanting fetal cells have been developing for more than 30 years in Sweden, Britain, and Mexico. Much of the medical research has concentrated on transplanting immature secretory cells into impaired adult brains and glands. Parkinson's disease and diabetes are targeted for the most research. Success with one disease offers greater hope for the use of fetal cell technology in the treatment of other diseases where new and healthy cells would be welcome—for example Alzheimer's, paralysis, AIDS, cancer, and aging.

In a typical fetal tissue transplant of brain cells, a patient will receive the cells by hypodermic injection, guided by tomography to the critical portion of the brain. The accepted and surviv-

Irwin L. Slesnick and Jal S. Parakh are professors in the biology department of Western Washington University, Bellingham, WA 98225.

ing embryonic cells grow normally in size and shape, extending axons and dendrites into the host brain. Soon, appropriate synapse connections will be made. In the case of Parkinson's disease, as the cells mature they secrete the neurotransmitter dopamine which is deficient in the system of these patients.

Issues

The use of fetal tissues in scientific and medical research is vigorously challenged by anti-abortionists. They believe that the harvesting of living cells from intentionally aborted fetuses is as outrageously immoral as the use of the victims of the holocaust in medical research by Nazi doctors. These opponents of fetal cell research perceive the acquired living cells of the unborn as poison fruit, and no matter how much good can be derived from their use, nothing can redeem the moral wrong of the induced abortion.

Advocates of fetal cell research argue that they are operating under a moral code based on compassion and rationality. They claim that the embryonic cells of fetuses are far too valuable to discard, and that to ignore this available resource would itself be an immoral act. Each year, 1.5 million fetuses are legally aborted at about the age their cells are most useful for research. In the U.S. alone, medical research with fetal cells offers hope for two million diabetics and Alzheimer's patients, 1.5 million victims of Parkinson's disease, and 300 000 people with spinal injuries. Against the argument of complicity in abortion, advocates of fetal tissue research argue that using these cells no more implicates the researcher in abortion than using the surviving organs from an automobile accident implicates one in the accident.

Beyond the issue of the morality of abortion, ethicists have been concerned about how society will regulate an industry based on the use of fetal cells, whatever their source. One projection envisages a $6 billion per-year medical business: As fetal cell transplants become routine, some fear the marketing of fetuses will take control of physicians and clinics. Or, women may be persuaded to become pregnant, schedule their abortion, and designate the person who is to receive their fetus' cells—or even sell their fetal cells in a personal business.

. . . these cells can integrate and mature with the tissues in the brain of an adult host.

During the latter part of 1988, an advisory panel on fetal tissue use in research recommended to the National Institute of Health by a vote of 17 to 4 that the U.S. government should continue to support fetal tissue research. In approving the continuation of research, the panel recommended the imposition of the following controls: A woman must have no monetary or noble incentives to have abortions or to delay the procedure in order to allow the fetus to develop; A woman must give her informed consent to the use of her fetus' cells only after she has made the decision to have the abortion; and, a woman may not direct the use of fetal cells nor should she be told anything about their destiny, in other words, anonymity must be maintained between donor and recipient.

Questions

You may find the questions on the accompanying student page sufficient to provoke discussion of which direction fetal cell research and treatment should take. The following questions could stimulate deeper thinking on this issue.

- Someone observed that transplanting parts of the dead into the bodies of the living is one of the strangest things humans have ever done. Will the security of being the person you were born as be affected by sharing your body with functional cells of an unborn fetus?
- When the National Institutes of Health suspended funding to 116 fetal tissue research projects, researchers protested that the action was dangerous to American interests. They argued that if the U.S. government didn't support this research, private companies and foreign countries would go forward without the safeguards NIH regulations ensure. How valid are such protests about money, regulation, and competition?
- A young woman's father has Alzheimer's disease. She is seeking help in becoming pregnant with her father's sperm so she can grow the most genetically compatible fetal cells for transplant. Should she and participating specialists be permitted to pursue this project?
- How are the arguments for or against fetal cell transplants similar to or different from those for or against transplants involving the heart, kidneys, or liver? ■

For further reading

Fine, A. (1988) "The Ethics of Fetal Tissue Transplants." Hastings Center Report. 18(3):5–8.

Mahowald, M.B., J. Silver, & R. A. Ratcheson. (1987) "The Ethical Options in Transplanting Fetal Tissue." Hastings Center Report. 17(2):9–15.

Weiss, R. (1988) "Forbidding Fruits of Fetal-Cell Research." Science News. 134(19):296–298.

Weiss, R. (1988) "Fetal-Cell Transplants Show Few Benefits." Science News. 134(21):324.

Should Fetal Tissue Research and Transplants be Permitted?

—*Art by Max-Karl Winkler*

Some day soon, the living cells of aborted fetuses may keep adults alive, healthy, and youthful. Fetal cells at the proper stage of development can reach maturity and function normally if transplanted into the organs of mature humans. People with such degenerative diseases as Parkinson's, paralysis, and insulin-dependent diabetes can replace their degenerative cells with new healthy cells by means of a fetal cell transplant operation. While the procedure is still in the experimental stages, early results with humans seem promising. As research and medical applications of fetal cell transplants go forward, philosophers have already started delving into the ethical issues involved in putting the living cells of aborted fetuses into the bodies of people for whom new tissue may mean longer, happier lives.

Today, the use of fetal tissue for any purpose arouses strong emotional responses, because of its relation to the grave moral issues associated with abortion, and the status of human embryos and fetuses. The questions that follow have provoked a considerable amount of controversy.

Questions to consider

1. Under what circumstances would you accept the cells of an aborted fetus to cure you of diabetes or a parent of Parkinson's disease?
2. Suppose the numbers of induced abortions in the U.S. declined dramatically due to the passage of anti-abortion legislation or the increased and effective use of birth control procedures. How might the fetal tissue industry meet its demand for fetuses? Should we buy and import fetuses from countries where induced abortion is legal?
3. What limits, if any, should be imposed on fetal tissue transplant research and development? Should fetal cells be injected into the facial skin of the aged who want to appear young with a rejuvenated face? Will the replacement of degenerated tissues and organs extend the human lifespan beyond the best interests of our species?
4. Will the medical use of fetal cells have brutalizing effects on our society that would outweigh the good of medical progress? Will our reverence for life erode a little more each time a aborted fetus is used to improve the life an adult?

—*Art by Max-Karl Winkler*

Difficult Decisions: The Superconducting Super Collider

by David E. Newton and Irwin L. Slesnick

The search for simplicity has been a recurring theme in the history of science. Scientists are continually looking for a handful of particles and forces that will explain the operation of the natural world. But, just when scientists think that they've discovered the "basic building blocks" of all matter, new evidence shows that those blocks can be divided into even smaller blocks. First, atoms were divided into protons, neutrons, and electrons; then the protons and neutrons were found to be composed of even smaller particles called quarks.

The theory that scientists use today to explain the composition of matter is called the Standard Model. This model assumes that two kinds of particles—quarks and leptons—are the basic building blocks of all matter. But scientists suspect that this theory may be incomplete. Thanks to research with particle accelerators, a new bewildering array of sub-atomic particles has been produced. Scientists have now found that different types of quarks exist at different energy levels. For example, two quarks, the "up" and "down" forms, and two leptons, the electron and the electron neutrino, can be found at the first energy level. Other particles that exist at this energy level are combinations of these particles. For example, the proton is a combination of two up quarks and one down quark.

At higher energy levels, such as those created by particle accelerators, new particles are found that do not exist at ordinary levels. Within the range of present-day accelerators, the "strange" and "charm" quarks and two additional leptons called the "muon" and the muon neutrino have been found. Many members of the particle zoo—the term used by physicists to describe the many subatomic particles discovered with accelerators—are merely combinations of higher level quarks. The omega particle, for example, has been shown to consist of three strange quarks, while the sigma zero particle consists of an up, a down, and a strange quark. (See Figure 1 for a summary of the fundamental particles of matter as predicted by the Standard Model.)

Another kind of elementary particle, according to the Standard Model, is that which transmits a force. The photon, for example, has long been recognized as a mediating particle for the electromagnetic force, that is, a "virtual" particle through which the force acts on other particles. The discovery of the W^+, W^-, and Z° bosons in 1983 was regarded as a powerful confirmation of the validity of the Standard Model. Figure 2 summarizes the four fundamental forces and the mediating particles associated with each.

Discoveries of this kind have prompted physicists to call for the construction of a superconducting super collider (SSC). Like earlier accelerators, this machine will propel protons, electrons, and other particles of matter at greatly accelerated speeds. Two beams of particles, accelerated in opposite directions, will be made to collide with each other and the resulting collision will reveal much about the composition of matter. Energy will also be produced. The energy of the earliest cyclotrons, for example, was measured in millions of electron volts (MeV). Modern day colliders, in contrast, typically produce energies measured in the billion electron volts (GeV) to trillion electron volts (TeV) range. The proposed SSC would generate about 20 TeV, nearly ten times the amount of energy presently available from the largest existing particle accelerator. Figure 3 summarizes the evolution of particle accelerators over the past half-century.

The decision to design a machine that would produce about 20 TeV was not accidental. That energy level appears to be the minimum necessary to answer some of the most interesting questions involving the Standard Model. For example, physicists want to know where particles get their mass. A critical factor in answering that question appears to be a particle hypothesized by Peter W. Higgs from the University of Edinburgh, the so-called Higgs boson. If that particle does exist, physicists believe that it should become apparent at energy levels close to 20 TeV. Similarly, evidence about the possible interrelationship between the strong and the electroweak forces may begin to manifest themselves within this energy range.

Policy issues of the SSC

The practical issue confronting legislators, scientists, and ordinary citizens is whether the SSC is worth its tremendous cost. Research to be conducted with the SSC falls into the category of "pure" or "basic" research. That is, the answers provided by the experiments will not solve any practical problems. They will only provide more information about the natural world. Information gained from the SSC will not lead to a cure for AIDS, make rockets travel faster, solve problems of infertility, or address any other issue of national significance.

Also, scientists are debating among themselves about whether the SSC is worthy of their support. Many worry that the money spent on the accelerator will be taken from research in biology, chemistry, medical science, or other fields of physics. Perhaps the general health of science would be better served, some argue, if the giant new machine were not built.

The "Questions to consider" on the following student's page will start students thinking about the issues surrounding the construction of the SSC. You may want to follow up these questions with additional problems such as the following.

- What are some of the arguments scientists offer in favor of and against the idea of building the SSC? How do the arguments of politicians and the average citizen differ from those of scientists?
- Proponents of the SSC argue that many practical benefits can be expected as spin-offs of building the machine. Critics say that these incidental benefits do not justify the cost of the machine. Explain which side you feel has the stronger argument.
- The selection in late 1988 of Texas as the site for the SSC promises to bring that state 2000 high-tech jobs with an annual payroll of about $270 million. Explain how this fact would affect your position about building the SSC if you were a Congressman from Texas, or a representative from some other state.
- You have been hired as a public relations writer by the committee designing the SSC. How would you try to convince a Congressional committee of the importance of discovering the Higgs boson? ■

David E. Newton is a free-lance science writer and adjunct professor at the University of San Francisco. Irwin L. Slesnick is a professor in the biology department at Western Washington University, Bellingham, WA 98225.

Figure 1. The fundamental particles.

Level	Quarks	Leptons	Energy level
3	top (or truth) bottom (or beauty)	tau tau neutrino	origin of the universe most powerful particle accelerator
2	strange charm	muon muon neutrino	cosmic ray events most particle accelerators
1	up down	electron electron neutrino	everyday objects and events

Figure 2. The fundamental forces.

Force	Relative strength	Acts on	Effective distance	Mediating particle(s)
Gravitational	10^{-39}	all particles	very long	graviton(?)
Weak	10^{-5}	leptons, quarks	very short	W^+, W^-, Z°, bosons
Electro-magnetic	10^{-2}	charged particles	very long	photon
Strong	1	quarks	very short	gluon

Figure 3. The evolution of particle accelerators.

Machine	Type*	Opened	Maximum energy (GeV)	Radius (Meters)
Early cyclotron (U. of Calif.)	c	1932	0.0012	0.25
Synchrocyclotron (U. of Calif.)	c/p	1946	3	4.67
Electron synchrotron (Cornell University)	c/e	1954	2.2	15
DESY electron synchrotron (Hamburg, West Germany)	c/e	1964	7.5	50
Stanford linear accelerator (California)	l/e	1966	50	3 km (length)
Proton synchrotron (Serpukhov, USSR)	c/p	1967	76	236
CERN proton synchrotron (Geneva, Switzerland)	c/p	1976	400	1100
Fermilab proton synchrotron (Illinois)	c/p	1972	1000	1000

*l = linerar accelerator; c = circular accelerator; e = electron bullets; p = proton bullets

The Complex Issue of Basic Research

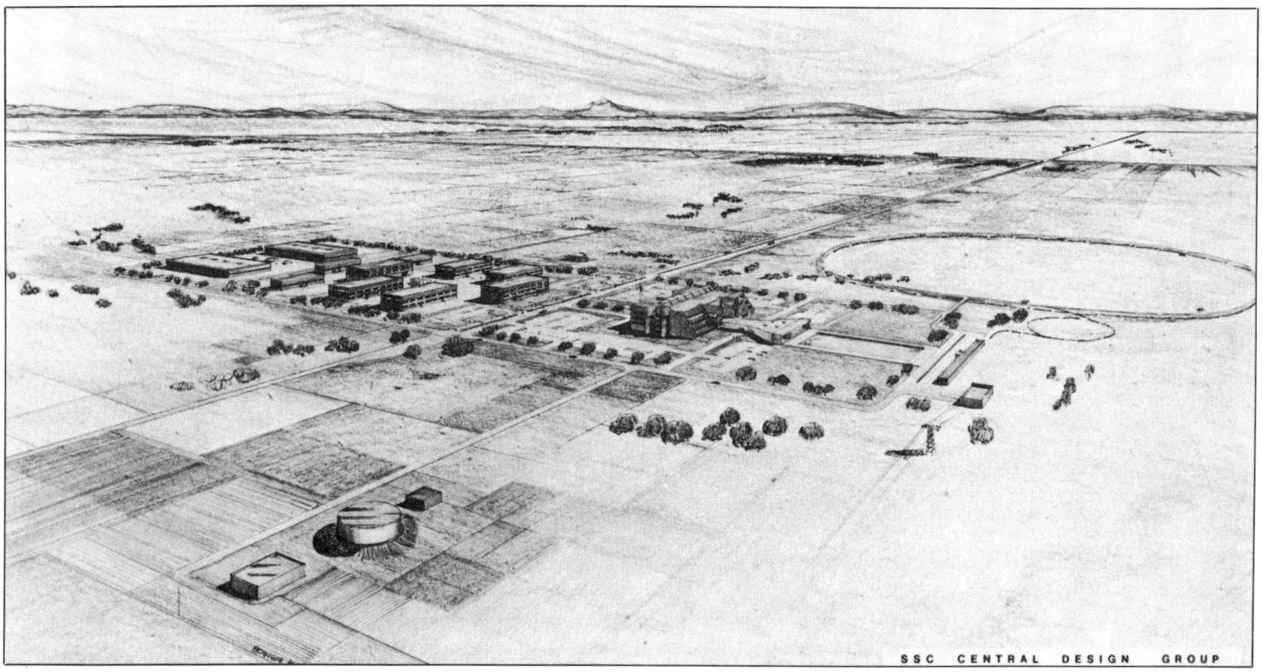

—*Art courtesy of the U.S. Department of Energy*

Would you pay $10 billion for this picture? Many physicists would like the American taxpayer to say "yes" to this question. They believe that machines such as the superconducting super collider (SSC) shown above will provide the next step in our understanding of the composition of matter.

Scientists have always believed that all matter consists of a small number of elementary particles, particles that cannot be broken down into anything simpler. That idea was expressed in 1803 in John Dalton's atomic theory. Dalton suggested that atoms were the smallest particles of which matter is composed. This idea was widely accepted for years, yet, it was inaccurate. As early as 1890, scientists began to realize that atoms were not indivisible and therefore not elementary particles.

Then, in the 1950s, scientists found ways to probe even deeper into atoms and discovered that still smaller particles exist. As it turned out, protons and neutrons were composed of even tinier particles called quarks. The key to discovering this fact was the invention of a powerful new research tool, the particle accelerator. These machines allowed scientists to speed up protons, electrons, and other particles of matter to very high velocities. The beams of particles generated by the accelerators made it possible for scientists to look deep into the structure of matter and reveal particles that had never been seen before.

Taking a peek at the particle zoo

In addition to quarks, accelerators revealed hundreds of other particles such as mesons, sigma particles, K particles, and J/psi particles. Scientists were astounded and confused at the new "particle zoo" that they had discovered. Which, if any, of these particles were truly elementary, they wondered, and which were compound particles like atoms and protons?

Eventually, physicists developed a new theory about the composition of matter, one that explained all the new particles created by accelerators. According to this Standard Model, only two kinds of particles are truly elementary: quarks and leptons. Every other particle we know about

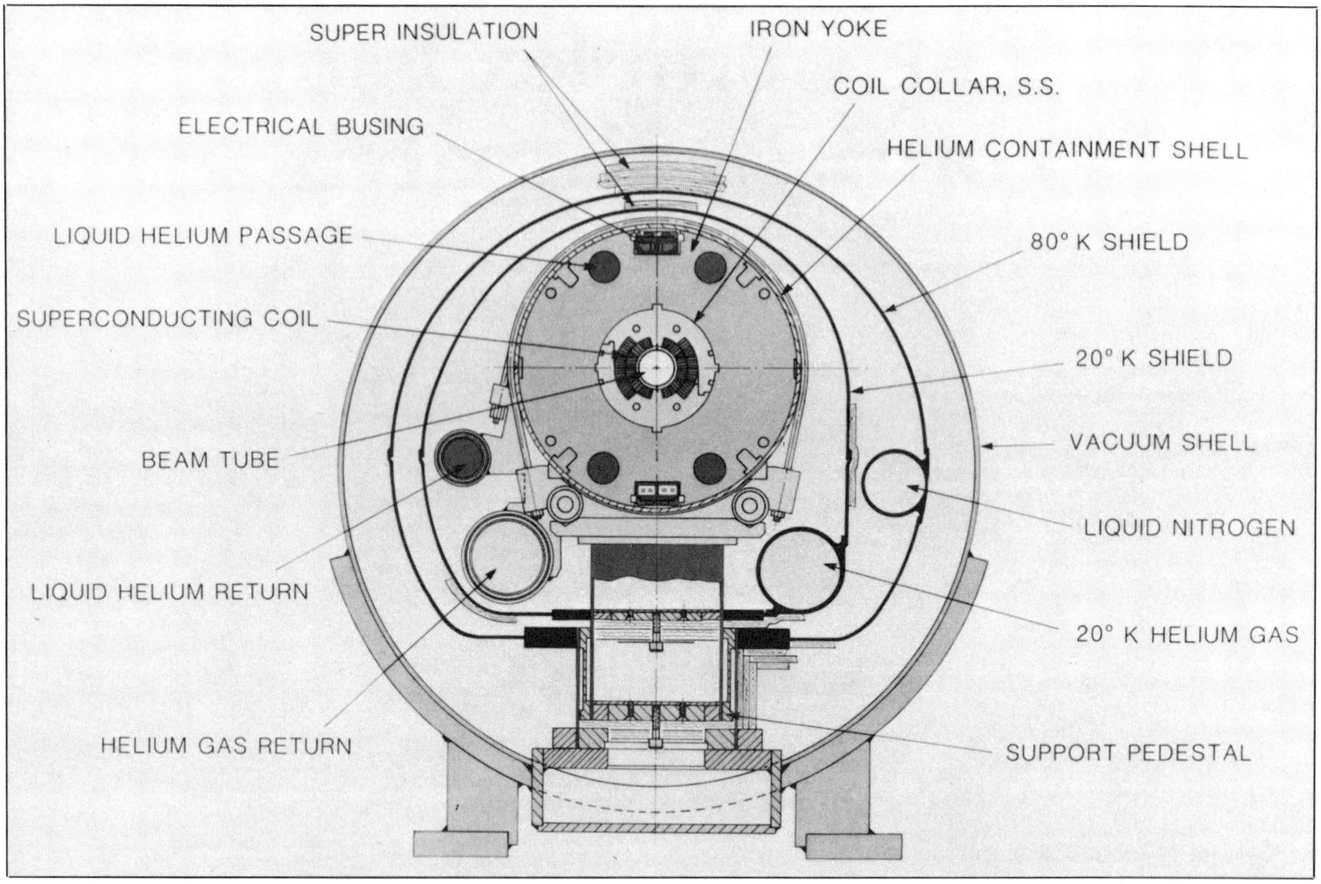

—Art courtesy of the U.S. Department of Energy

consists of some combination of the two kinds of fundamental particles.

The Standard Model has answered some basic questions in physics during the last decade, but it is still incomplete. For example, many physicists now think that quarks and leptons aren't really elementary but consist of even smaller particles. The problem is that no existing particle accelerator is powerful enough to answer any of the questions brought up by this new theory. Particle physicists are limited in the research they can carry out because they cannot obtain any new data for their theories about the composition of matter.

Thus, for nearly 10 years, particle physicists have been working on plans to build the most powerful accelerator ever conceived, the superconducting super collider (SSC). This machine would consist of a hollow tube about 3 m in diameter and 83 km in circumference. Beams of protons traveling in opposite directions would collide within the tube generating 20 trillion electron volts (TeV) of energy. The colliding beams would produce particles never before seen by humans, particles that should answer important questions about the Standard Model.

The SSC would be a very expensive machine, however. It is expected to cost at least $6 billion to build and at least $250 million to operate each year. Some scientists think that this is too much to spend on a single project. They fear that the SSC will take funds away from other kinds of scientific research. Many people object to the SSC because they see no practical benefit arising from it. They point out that the new accelerator will contribute nothing to improving human health, solving medical problems, making life easier or providing a stronger defense.

Questions to consider

1. Had you been a member of the U.S. Congress in 1988, would you have voted to approve of President Reagan's request for $363 million to start work on the SSC? Why or why not?
2. Some people argue that we should build the SSC in order to keep the United States a leader in particle physics research. Explain why you agree or disagree with this argument.
3. Many research projects funded by the U.S. government are designed to answer questions about the natural world, but have no apparent value. To what extent, if at all, should the government finance research of this kind?

Difficult Decisions: Should Modern Man Hunt?

by John A. Miller and Irwin Slesnick

Evidence from the fossil record suggests that humans have been hunters for about 3 million years. *The Hunting Hypothesis,* by R. Ardrey, describes our ancestors as meat-eating apes who hunted in packs like wolves, using lethal weapons. Over 99 percent of human history was hunting-dependent. Cultures attached a high value to hunting skills because by natural selection, those who were best at hunting were also the most likely to survive.

In the nineteenth century, hunting turned into an ugly enterprise. No longer was it a primary method of obtaining food or even a sport; hunting became a business to control the world food market. In the process, market hunters disrupted natural systems by destroying large carnivores, such as mountain lions and wolves. They also killed off herds of bison to subjugate the Indians and so that high-yield crops and livestock could replace wildlife.

At the beginning of the twentieth century, when the populations of many species of game animals were at an all-time low, many sportsmen who loved to hunt realized they must accept the stewardship for wildlife in the United States or lose their heritage. The sport-hunting public, led by Teddy Roosevelt, Aldo Leopold, and John James Audubon, initiated a conservation effort that included passing laws to save many species of wildlife from being totally lost to future generations.

A series of acts, including the Lacey (1900), the Migratory Bird Treaty (1918), the Duck Stamp (1934), the Pittman-Robertson (1937), and the Dingelt Johnson (fishing 1950), placed nonmigratory game management under the control of the states, eliminating market hunting for most species. Income from licenses and migratory waterfowl stamps provided a funding base that now exceeds $450 million per year. The purchase of land designated for wildlife and the development of scientifically based game management was intended to benefit both game and other wildlife.

Today, hunting remains closely linked to game management. Many species are managed specifically to provide quality hunting. Some species (such as the ring-necked pheasant and partridge) have been added to existing habitats for hunters, and others (such as wild turkeys) have been restored in areas where they had once disappeared. Major portions of the budgets of state natural resource departments come from hunters and go to management of game.

Modern hunting is embroiled in a major controversy. There are approximately 17,500,000 hunters confronting an indeterminate number of animal rights activists who are attempting to have hunting banned in the United States. The two groups are polarized. Confrontations are occurring in the courts, the state legislatures, the U.S. Congress, and in the field. Those opposed to hunting challenge the rights of people to kill game animals, the funding base, and the use of hunting as a game management tool. However, the core of their concern with hunting is the morality of the sport.

Are humans hunters by nature?

Yes, say pro-hunting sportspersons. Hunting is our heritage and an integral part of our interaction with nature, leading to our enjoyment of the outdoors. Over 17 million Americans seek this type of challenge, and attempt to learn the complex relationships in nature that affect hunting success.

Anti-hunting activists disagree. While humans probably evolved for several hundred thousand generations when hunting skills were necessary for survival, the activists say that the perpetuation of ancestral behaviors isn't justified. Modern humans certainly do not condone warfare, cannibalism, or suppression of women just because the behaviors existed in the past.

Is hunting a necessary game management tool?

Anti-hunting people say no. They deny that hunting is required for wildlife management. Many game animals undergo natural fluctuations in population size without any interference from humans. The hunter is accused of killing the genetically strong rather than the less fit, and thus changing the gene pool of the species.

Hunters disagree. They argue that nature must provide a surplus to ensure survival of the species, and as many as 80 percent die each year. Furthermore, in the case of deer and elk, the habitats in which they live have often changed dramatically. For example, reduction in woodland has resulted in far more deer in the upper mid-West than in the previous century. They deny that hunters exert a reverse-selection pressure, since they normally shoot the first animal they can find, which may often be a genetically inferior one. They also cite experiments at other types of population control, such as in the Everglades, which were not successful. Hunters also argue that it is better for an animal to die from a quick shot than from starvation.

—*Art by Max-Karl Winkler*

Do hunters control the refuges?

National wildlife refuges were originally set up as "inviolate sanctuaries" for all living species. While duck stamp revenues do support management programs there, more than 95 percent of the 89 million acres of refuge lands was purchased with tax dollars derived from sources unrelated to hunting. Since Congress opened large areas of our Wildlife Refuge System to hunting in the 1960s, pro-hunting sources claim that improved management of many species of game has been documented.

Yet hunters make up only 3 percent of refuge visitors, and to those opposed to hunting, their control of the uses of refuge lands is perceived as excessively self-serving and unnecessary.

Is hunting moral?

There are many who insist that caring, civilized people should not hunt sentient animals for fun, trophy, or sport because of the trauma, suffering, and death that result and that a society that characterizes wild animals as "game" denies respect for life.

The pro-hunting faction finds the argument that it is immoral to kill wild game, while it is morally acceptable to destroy their habitat in order to raise crops and livestock for slaughter, to be inconsistent. In contrast they say that hunters are thoughtful predators, who feel sadness at taking a life but still accept the natural role that humans have in nature.

Do anti-hunting persons have the right to deny hunters their right to hunt?

Most hunters no longer have an economic need to hunt. Quick, clean slaughters by professional butchers now provide healthy meat both economically and efficiently. Therefore anti-hunting activists feel they have both the right and the duty to pursue legislation to protect wildlife and ecosystems from violence and disruption.

In contrast, the hunter does not believe that one group of citizens has a right in a free society to deny another group their heritage and historical pursuit of happiness when it does not impinge on the freedom of others. They call the activist claims that hunters are interfering with natural ecosystems unsubstantiated, and instead insist that it was the hunters who accepted responsibility to protect wildlife at the beginning of this century and continue to support management today. ■

For further reading

Ardrey, R. (1976)*The Hunting Hypothesis.*1st ed. New York,: H. Wolff.

Conley, C. (1988) "Chip Shots and the Outdoors." *Outdoor Life,* 182 (5): 6.

Favre, D. S. and G. Olson (1982) "Surplus Population: A Fallacious Basis for Sport Hunting." *Society for Animal Rights,* 421 S. State St., Clarks Summit, PA 18411.

Reiger, G. (1986) "Not by Bread Alone." *Field and Stream,* 91(1): 26, 28.

Teague, R. D. Ed. *A Manual for Wildlife Conservation.* Washington, D.C.: The Wildlife Society, 1971.

Willis, H. (1989) "Unwelcome Whitetails." *Field and Stream,* 93(10): 58-59, 89-90.

Note

For various and continuing publications on hunting contact the Humane Society of the United States, 2100 L St. NW, Washington, DC 20037, (202) 452–1100 or Wildlife Conservation Fund of America, 50 West Broad St., Columbus, OH 43215, (614) 221–2684.

To Hunt or Not to Hunt

lunters' view
lunting is a sport for men and women who use ieir historical rights to stalk and shoot game for ood and trophy as part of a cherished outdoor xperience.

Vildlife management was begun by such hunters as 'eddy Roosevelt and John Audubon. Contrary to vhat anti-hunting activists claim, game management currently and primarily supported by hunters' cense fees and special taxes on arms and ammuni-on. Since conservation minded hunters have sup-orted wildlife programs, the numbers of wild ani-ials have increased.

1 the natural state populations of wildlife tend to icrease and decrease in cycles. Humans, however, ave changed so much of the natural environment or our own use that the natural cycles for wildlife re no longer a constant. Thus animals, such as deer, ften die slowly in agony from hunger, cold, or dis-ase often caused by overpopulation of a restricted nvironment. Hunters harvest animals under the ontrol of wildlife scientists. Tranquilizer guns are otally ineffective for hunting because of their short inge.

Aany hunters are concerned citizens who work for ie maintenance of abundant and healthy wildlife. In free society, their historical right to hunt must ever be denied by anti-hunting activists with dif-erent moral values.

Anti-hunting activists' view
Sports hunting of wildlife is a recreational pastime where hunters derive pleasure from stalking and shooting mammals and birds who are sensitive to pain and have a right to live in the natural environment.

Most funds for the purchase and maintenance of wilderness land come from general taxes. The payment of special taxes and the purchase of duck stamps do not of themselves carry rights to hunt on land purchased for wildlife refuges.

Population fluctuation has always been common in species. Natural populations do not require humans to stabilize their sizes. When hunters harvest small animals, they usually go for trophy specimens—the genetically strong rather than the small and weak. When necessary, excess populations should be harvested by professional biologists using painless tranquilizers and lethal injections.

Hunting is a violent blood sport that has lost its original purpose, and now serves only to entertain hunters and kill and wound the fittest wildlife.

Questions to consider

1. When game management fails to produce stable opulations of such wild animals as deer, should overnment hunters harvest the surplus rather than vite sport hunters onto the refuge to track and ioot the animal?
2. To what extent should the law permit an anti-unting activist to disrupt a hunter in the field?
3. Is there a moral difference between killing a ild animal for food and raising an animal in captiv-y for slaughter?
4. Is it fair or reasonable to ban hunting, thereby causing loss of funding for game management, funds for habitat purchase, and maintenance plus economic hardship for supply and sporting equipment companies?
5. What kind of compromise would you suggest for the hunter and the anti-hunting activist?
6. Can a person have a reverence for nature and wild animals even though he or she hunts?

Difficult Decisions: Animal Rights

Who has the right to determine animal rights?

by Jal S. Parakh and Irwin L. Slesnick

Since the scientific revolution started about 500 years ago, the use of animals in scientific research has been accelerating. Currently, millions of animals (nonhumans) are used for research each year. This research includes both medical experimentation and the toxicity testing of thousands of compounds, ranging from pesticides to household products and cosmetics.

Opposition to animal experimentation has been increasing too. Some antivivisection groups advocate stopping all animal experimentation. At the other extreme lie researchers who would allow any animal experimentation in pursuit of new knowledge. The conflict between these two groups, and their less extreme colleagues, has led to bitter name-calling, protests, demonstrations, and even break-ins at animal research facilities.

In recent years, the vivisection debate has centered on the issue of whether all sentient vertebrates have rights. Philosophers, scientists, lawyers, and other thinkers have been struggling to decide what kinds of rights animals might have and what the source of these rights could be. They have described a number of ethical positions on this issue, including absolute dominionism, abolitionism, and utilitarianism (Tannenbaum and Rownan, 1985).

Absolute dominionists deny that nonhuman animals have any rights. They maintain that this lack of rights means that there are no moral constraints on the use of animals for even trivial human purposes. In support of their stand, dominionists often cite a passage from Genesis stating that God gave humans dominion over all creatures. Opponents of this argument often point out that the absolute dominionists are using this passage selectively and neglecting a passage on stewardship of animals that also appears in Genesis.

Abolitionists assert that the use of animals for research is morally wrong because animals have the same rights as humans. It follows that if humans have a right not to be harmed, then so do other animals. Peter Singer, author of the provocative book *Animal Liberation,* labels prejudice in favor of humankind as "speciesism," a position he finds no more acceptable than racism.

Opponents of abolitionism argue that rights are seldom absolute and that when the rights of humans and other animals compete, human rights take precedence. One prominent abolitionist, Tom Regan, was forced to admit that in a "lifeboat situation," a dog should be thrown overboard before a human.

The most common justification for animal research is based on the ethical theory called *utilitarianism.* Utilitarians acknowledge that animals have rights, but they justify animal research by its outcome, saying that research gives us the best possible balance of good over bad consequences. In other words, the pain and death of nonhuman animals in research is outweighed by the research-related reduction in death, disease, and suffering of humans and other animals. A major problem with the utilitarian view is setting the standards by which we weigh the good and the bad.

Questions for discussion

You may find the questions on the accompanying student page (see p. 37) sufficient to provoke discussion about animal research. The following questions could stimulate deeper thinking on this issue.

- The ability to reason, to communicate, and to be conscious of one's mortality are among the differences between humans and other animals that often are cited to deny claims that nonhuman animals can have rights. If these criteria are to determine possession of rights, then what rights are held by fetuses, newborns, and people in irreversible comas?
- Reformers and animal protection advocates often propose the following guidelines: one, only those experiments should be permitted that will lead to a significant increase in our knowledge; two, that repeating the same experiments with different species of animals is unnecessary and should not be permitted; and three, that alternatives to animals, such as computer simulations and tissue cultures, should be used to test drugs and other chemicals. To what extent are the above restrictions realistic in terms of the nature of science?
- The Draize test, which involves applying cosmetics and household products to the eyes of rabbits, and the LD 50 test, which involves finding the dose of a substance that is lethal to 50 percent of the animals exposed to it, are both controversial on scientific and humane grounds. Have students investigate the arguments for and against these tests.
- Should all living things—bacteria, plants, worms, mammals—be equally protected from the "abuse" of scientific research? Do all life forms have equal rights? ■

Jal S. Parakh and Irwin L. Slesnick are professors in the biology department of Western Washington University, Bellingham, WA 98225.

References

Tannenbaum, J., and A.N. Rownan. (1985) "Rethinking the Morality of Animal Research." *Hastings Center Report,* 15(5):32–43.

For further reading

Fox, M.A. (1986) *The Case for Animal Experimentation: An Evolutionary and Ethical Perspective.* California: University of California Press.

Regan, T. (1983) *The Case for Animal Rights.* California: University of California Press.

Singer, P. (1975) *Animal Liberation: A New Ethics for Our Treatment of Animals.* New York: Random House.

Do We Have the Right to Do Research on Animals?

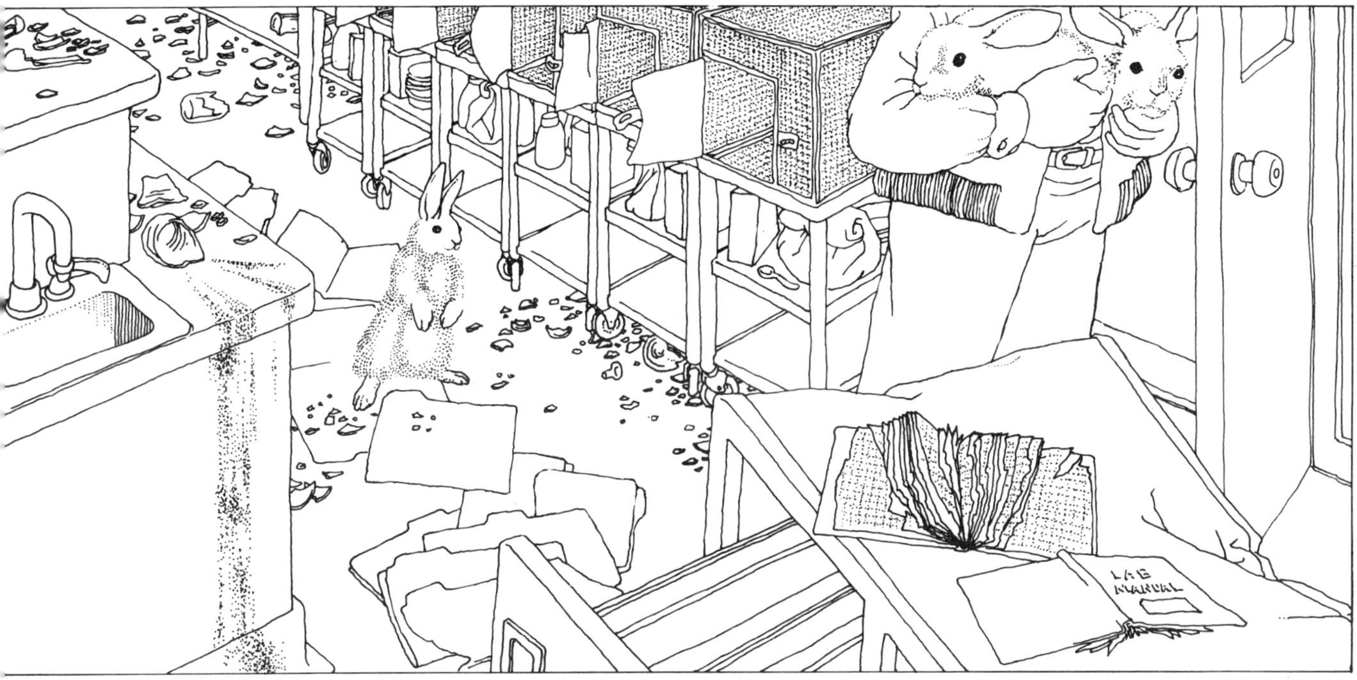

—*Art by Max-Karl Winkler*

"Animal Rights Activists Break in Lab, Free Crippled Cats . . . Teenage Protester Claims Animals Have Rights Too." Have you ever seen a headline like this one? Protests in support of "animal rights" are becoming more and more common.

Scientific and medical researchers routinely use animals in experiments. The researchers often have to expose animals to procedures that are painful and cruel by human standards. Sometimes, animals are killed in the course of experiments.

Opponents of animal research say that animals have just as much many rights as humans to be free of pain, abuse, and exploitation. Medical researchers point out that many medical advances came about through animal research. They are afraid that without the opportunity to experiment on animals, scientists will not be able to find cures for many diseases. And government regulatory agencies and manufacturers of drugs, cosmetics, and household products worry that without animal tests, harmful products could be put on the market.

What do you think? Should we allow animals to be used in research?

Questions to consider

1. Is there a difference between using animals for food or clothing and using animals for research?
2. Should a medical research center be allowed to use thousands of test animals in research on plastic surgery? What about research on a fatal disease?
3. Some research involves little or no pain to animals, or any shortening of their life spans. Other research involves significant animal pain and death. Should we consider these factors when deciding whether to allow certain research projects, or should we only look at the possible beneficial outcomes of the experiments?
4. Do researchers have the right to experiment on animals for the sake of knowledge alone? For example, do scientists interested in brain anatomy have the right to operate on cats' brains, causing the cats permanent disabilities?
5. Do cosmetics manufacturers have the right to test the safety of a new eyeshadow or lotion by placing samples of these products in rabbit's eyes? Do you think new products should be sold without first undergoing safety tests?
6. Does it matter what kinds of animals are used in animal research? Is it "more alright" to use a worm or other invertebrate, than to use vertebrates such as rabbits and dogs?

Difficult Decisions: Acid Rain

by John A. Miller and Irwin L. Slesnick

Rain, rain, go away; your pH is too low today.

Acid rain is a term used to describe any form of precipitation with an unnaturally low pH. If the pH of acid rain or snow is less than 5, the precipitation can severely damage forests and aquatic environments. The primary component of acid rain is sulfuric acid, created by the reaction of sulfur dioxide with atmospheric water. The primary industrial source of sulfur dioxide is emissions from factories and power plants.

The series of equations below illustrates the reactions sulfur dioxide undergoes as it forms sulfuric acid.

$$SO_2 + H_2O \rightarrow \underset{\text{sulfurous acid}}{HSO_3^- + H^+}$$

$$2HSO_3^- + O_2 \rightarrow S_2O_7^{2-} + H_2O$$

$$S_2O_7^{2-} + H_2O \rightarrow \underset{\text{sulfuric acid}}{2SO_4^{2-} + 2H^+}$$

This series of reactions proceeds over a 3- to 4-day period, so the acid often precipitates far downwind of the site where the sulfur dioxide was emitted. This phenomenon has had strong political implications. (Nitric acid also contributes to acid rain, but it is generally a localized problem.)

Acid rain that falls directly into lakes or streams will increase the acidity of the water gradually. In the case of snow, the rapid springtime melting of a winter's accumulation can cause an acid pulse, or a sudden, sharp decrease in pH, to occur. The effect the acid has on aquatic life depends on the chemistry of the body of water. If the lake or stream has a limestone bottom, very little damage occurs. The calcium and magnesium carbonates that make up limestone neutralize the acid.

On the other hand, nonlimestone substrates, such as clay and granite, do not contain carbonates and do not have the capacity to neutralize acid. As a result, pH levels in bodies of water lined with these materials drop steeply in the presence of acid precipitation. Low pHs (below 5) cause aluminum-based clay and granite compounds to break down and release aluminum ions. Both the free ions and aluminum hydroxide are harmful to trout. In addition, the low calcium levels in acidic lakes and streams may cause bone malformations in trout. Finally, trout food species, such as

John A. Miller is a chemistry professor and Irwin L. Slesnick is a biology professor at Western Washington State University, Bellingham, WA 98225.

—Photo courtesy of Soil Conservation Service, USDA

mayflies and dragonflies, do not reproduce at low pHs.

The most severe loss of trout habitats in the United States and Canada has occurred in the northeastern part of North America. Many of the high mountain lakes of the Adirondacks have lost over 40 percent of their original alkalinity and some can no longer support trout. Birds also may suffer ill effects from eating aquatic insects that are contaminated by the increased levels of aluminum present in the acidic waters. Furthermore, acid deposition can cause reduced plant growth and reproduction in forests.

Since Canada has an enormous economic stake in its forestry products, industry, and fishing-related tourism, acid rain emerged as a major issue between Canada and the United States during the late 1970s. The United States is the largest contributor to acid rain on this continent, producing 84 percent of the sulfur dioxide emissions. Even though the technology exists to reduce sulfur dioxide significantly (70 to 90 percent) by use of fluidized bed combustion or limestone injection multistage burners that can be installed in existing furnaces, U.S. industry has made little progress reducing acid precipitation. Ohio and other states with a decreasing industrial base are very reluctant to take any steps that would cause plant closings or reduce the sale of high-sulfur coal. Thus, most of the protests about acid rain tend to be localized in areas that feel the brunt of the problem, such as the northeastern United States and Canada.

Strained international relations caused by acid rain are not limited to North America. Norway and Sweden have experienced similar acid pollution in their lakes, which they blame primarily on Britain and, to some degree, on West Germany. However, the British argue that at least 45 percent of the offending sulfur dioxide cannot be traced, as it is part of the large mass of air that enters the Scandinavian countries after circulating over the Atlantic Ocean for a period of time. Thus, Britain is reluctant to pay dearly for cleaning its air when the origin of the acid is not definitely known and is possibly multinational.

Questions for discussion

You may find the questions on the accompanying student page (see page 34) sufficient to provoke discussion about acid rain. The following questions could stimulate deeper thinking on the issue.

- In some nations, including the United States, 90 percent or more of energy consumption originates from fossil fuels. In other nations, a higher percentage of consumed energy originates from nuclear, hydroelectric, and geothermal sources. Should the U.S. replace its fossil-fuel burning plants with nuclear-powered plants? What would be the consequences of such a switch?
- Rewriting our air pollution standards to reduce sulfur dioxide emissions significantly could cause serious economic consequences. Is it right that pollution control should cause a company to go bankrupt or be less competitive in the world market? Or cause stockholders to earn less dividends and workers to have reduced wages?
- Should Canada be able to sue a factory owned by an American company, but located in Canada, to recover all acid rain damages caused by emissions of the American-based plants of the same company? ■

For further reading

Chang, S.G., D. Littlejohn, and K.Y. Hu. (1987) "Disulfate Ion as an Intermediate to Sulfuric Acid in Acid Rain Formation." *Science*, 237 (4816):756–758.

Dickman, S. (1987) "Canada Demands Action From United States on Acid Rain." *Nature*, 325(6105):565.

Lucoma, J.R. (1984) *Troubled Skies, Troubled Water, The Story of Acid Rain.* New York: Viking Press:

Schindler, D.W. (1988) "The Effect of Acid Rain on Freshwater Ecosystems." *Science*, 239 (4836):149–157.

Stillwell, N. (1984) "Our Trees Are Dying." *Science Digest*, 92(9):39–48,96.

Who Should Pay to Clean Up Acid Rain?

—*Art by Marilyn Miller*

For many years, your family has been going trout fishing in the Algonquin Highlands of Canada. Each year, the fishing has been more and more disappointing. The trout are fewer and scrawnier and the mayfly larvae you use as bait are getting smaller and smaller. The Canadians who live nearby say that acid rain from the United States is responsible.

Many scientists agree with the Canadians. They think that acid rain is endangering fish and forests in Canada, the northeastern United States, and Scandinavia. They blame the acid rain on sulfur dioxide, a pollutant released from the smokestacks of coal- and oil-burning power plants and certain factories, such as paper mills and metal smelters. The sulfur dioxide combines with water and oxygen in the atmosphere to form sulfuric acid. Rain and snow wash the sulfuric acid out of the sky, into the lakes, streams, and forests below. Often, the acid rain falls far downwind of the factory or power plant that produced it. That's why the Canadians claim their acid rain problem is from the United States.

Smokestacks can be fitted with special devices that remove sulfur dioxide from exhaust. But these devices, including scrubbers, burners, and other combustion equipment, are costly. Installing and using them can raise electricity and manufacturing costs by as much as 15 percent. Since acid rain often drifts downwind, who should pay for the equipment to save the life in lakes, streams, and forests?

Questions to consider

1. Should people who live in Ohio pay for scrubbers and combustion equipment in Ohio factories to save the fish and forests of New England?
2. Should Americans pay to clean up Canadian lakes? Should Canadians help pay for scrubbers in the United States? Should the costs of fishing licenses be increased to pay the antipollution costs?
3. Should the United Nations, or some other international authority, regulate the industrial emissions of all nations in the interest of clean global air?
4. Should the U.S. government require all factories and power plants that produce sulfur dioxide to purchase scrubbers and combustion equipment?

Difficult Decisions: Chemical Warfare

by Irwin L. Slesnick and John A. Miller

From LSD to agent VX—the pros and cons of chemical weapons.

Chemical warfare probably began in prehistoric times with the use of poisoned arrows. Later, during the siege of Athens in about 400 BC, the Athenians were attacked by the Spartans with irritating sulfur dioxide that they produced by burning sulfur with pitch upwind of the city.

World War I was the beginning of modern-day chemical warfare. In 1914, the French lobbed the first tear gas grenades; the Germans retaliated with tear gas in artillery shells. On April 22 of the following year, the Germans used chlorine gas in Belgium, killing about 5000 Allied soldiers. The Germans released the first mustard gas, $S(CH_2CH_2Cl)_2$, and phosgene, $COCl_2$, against the Allies in 1917 at Verdun. By the end of World War I, about 3000 different toxic chemicals had been considered for use in war.

GA

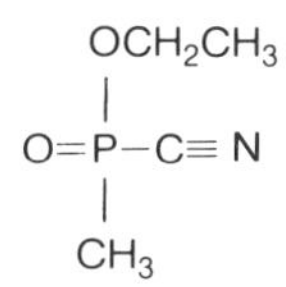

—Art by Max-Karl Winkler

The 1925 Geneva protocols outlawed first-strike use of chemical and biological warfare. (The United States ratified this treaty in 1974.) Even so, since the Geneva protocols, gas has been used by Italians against Ethiopians, Japanese against Chinese, Egyptians against Yemenites, and Iraqis against Iranians. Although no gases are known to have been used in World War II, the Germans did test nerve gases on concentration camp prisoners.

During the Vietnam War, the United States used nonlethal riot control gases in tactical situations, claiming that these harassing types of agents do not qualify as war gases prohibited by the Geneva protocols.

Antipersonnel chemical weapons can be grouped according to their physiological effects. Nerve agents are the most lethal chemical munitions for tactical or strategic warfare. These organophosphorus compounds interfere with the proper transmission of nerve impulses, by irreversibly inhibiting acetylcholinesterase, an enzyme crucial to the operation of the nervous system.

There are two classes of nerve gases—the earlier, less toxic G agents and the modern, highly toxic V agents. The first of the G agents was GA, with the trivial name Tabun and the formula shown at left. Agent VX is probably the most deadly form of war gas, because of the small dosage necessary to kill. But there is no reason to believe that the limit of lethality of nerve gases has been reached.

Chemical agents of lower lethality than nerve gases include blistering agents, choking agents, and blood agents. This category includes the blistering agent mustard gas and the choking agent phosgene. As a blood agent, hydrogen cyanide affects the absorption of oxygen by hemoglobin. These chemicals can cause great pain and, often, lingering death.

Of lowest lethality are the riot control agents and the relatively new incapacitating agents. Riot control agents can cause such reactions as tearing, sneezing, itching, and vomiting. Chloroacetophenone, shown at right, is the tear gas known as Mace™. Adamsite™ is the vomit gas diphenylaminochloroarsine, also shown. The dosage of these agents used to disperse rioters is much less than that which is lethal, but even such a dosage can be lethal to victims in confined spaces or who are sick, young, or old.

H. G. Wells introduced the concept of incapacitating agents in *Things to Come* (1935), in which two warring nations were locked in a hopelessly suicidal conflict. The good guys gassed the weary warriors on both sides with an anesthetic that induced narcosis long enough for disarmament. Physical incapacitants are attractive nonlethal weapons because, as proposed by H. G. Wells, they promise humane warfare with neither pain nor death. They physically immobilize people by inducing sleep, paralyzing skeletal muscles, or making it impossible to remain standing without fainting.

Military expenditures for chemical warfare are high—more than $1 billion per year in the United States alone. As early as the 1950s, 20 percent of our chemical warfare budget went toward research and development of physical incapacitants. At present, the United States is believed to possess only one incapacitating agent, BZ. Presumably, this agent is a reversible inhibitor of acetylcholinesterase, in contrast to lethal nerve gas, which is an irreversible inhibitor. Research findings and combat strategies have been and still are top military secrets, so the practicality of bloodless combat with incapacitating agents is unknown.

One of the most serious problems with using physical incapacitants as

Irwin L. Slesnick is a professor in the biology department and John A. Miller is a professor in the chemistry department at Western Washington University, Bellingham, WA 98225.

chloroacetophenone

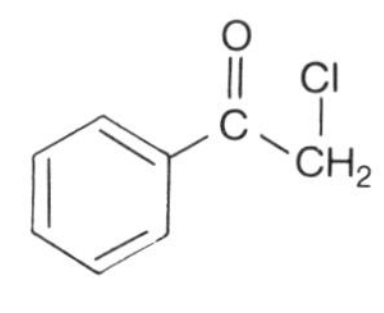

diphenylaminochloroarsine

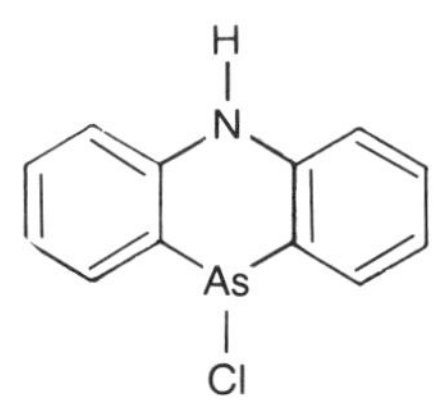

strategic weapons is that their dosages may range into lethality for general populations. That is, the quantities of gases dropped on civilian targets can be adjusted for normal adults. But such doses can be fatal to children, the elderly, or the sick. This problem is compounded because weapons sprayed into the air cannot be aimed and fired like bullets—gases and smokes drift with the changing winds.

Testing incapacitating agents presents another dilemma. Researchers cannot assess the usefulness of a chemical agent without tests on human subjects. Chemicals that only incapacitate test animals such as rabbits, goats, and monkeys might have different effects on human test subjects, perhaps severely injuring or killing them. So, who could ethically be tested?

Questions for discussion

You may find the questions on the accompanying student page (see page 33) sufficient to provoke discussion about the wisdom of using nonlethal physical incapacitating agents in war and in police work. The following questions could stimulate deeper thinking on this issue:

- The Germans discovered and produced nerve gas in the 1930s. They tested and used it on prisoners in concentration camps. They never used it against the Soviets or the British, even when their victory was threatened. Why do you think they did not use their nerve gas? (Hitler didn't use his stockpile of nerve gases not out of moral considerations but for fear that the Allies also had G agents with which they would retaliate.)
- Soldiers were ordered to attend the testing of nuclear weapons in the 1950s. Do you think the military should have the right to ask a soldier to volunteer, or to order a soldier, to participate in a test of an incapacitating agent? Do you think that prison inmates should be offered reductions of sentences, or commutation of death sentences, in exchange for participating in tests of incapacitating agents? Why do you think as you do?
- Suppose that during World War II either Japan or the United States (but not both) had had highly effective nonlethal incapacitating agents. How might the war have been different? ■

For further reading

"Bad Chemistry." *The New Republic*, 190:7–9, May 21, 1984.

Cookson, J., and J. Nottingham. *A Survey of Chemical and Biological Warfare*. New York: Monthly Review Press, 1969.

Hersh, S. *Chemical and Biological Warfare*. Indianapolis: Bobbs-Merrill Company, 1968.

Morrisey, D.H. "The Return of Chemical Warfare." *The Progressive*, 46:25–28, February 1982.

Stockholm International Peace Research Institute. *The Problem of Chemical and Biological Warfare*, Vols. 1–6. New York: Humanities Press, 1971.

Can New Chemical Weapons Lead to Humane Warfare?

—Art by Max-Karl Winkler

Nearly all modern warfare is chemical. Most weapons depend on chemical reactions to burn and blast. But society usually uses the term *chemical warfare* to refer only to the intentional exposure of people to toxic substances that can kill, sicken, or incapacitate them.

Thousands of different compounds have been developed and tested for their toxicity. The most deadly of all is the nerve gas VX. A drop of VX the size of the period at the end of this sentence is fatal. Nerve gases kill people the same way Raid™ kills roaches, by blocking the proper transmission of nerve impulses.

Blistering and choking gases, such as mustard gas and phosgene, are less deadly than nerve gases. But they cause severe irritation of lungs, eyes, and skin. Each victim requires help from several other soldiers. The chaos resulting from mustard gas or phosgene attacks might shorten a war . . . or invite retaliation.

In recent years, chemical warfare researchers have been developing two categories of nonlethal chemical agents that incapacitate people. "Off the rocker" agents, such as LSD–25, can neutralize a population by making victims incapable of realizing what they are doing for hours. Military strategists have nearly abandoned these psychologically incapacitating agents out of fear that mass madness might panic bewildered leaders into nuclear war.

"On the floor" agents cause people to collapse by lowering their blood pressure, putting them to sleep, or blocking the operation of their involuntary muscles. A highly secret chemical, known as BZ, may act on the nervous system in a way similar to that of nerve gas. BZ is thought to be a true "on the floor" incapacitant, as opposed to disabling chemicals such as Mace™. Mace is a highly purified form of tear gas, a riot control agent. It severely irritates nerve endings in eyes and mucus membranes, making those sprayed with it very uncomfortable. But, while Mace is a nonlethal chemical agent, it is not a true incapacitant.

Questions to consider

1. Do you think our military should be conducting research to develop nonlethal physically incapacitating agents? Why do you think as you do?

2. Assume that decades of high-priority research have produced incapacitating agents that can immobilize people as effectively as nerve gases can kill them. Under what conditions should these incapacitating agents be used? Should they be used strategically (against an enemy's ability to make war)? Tactically (on the battlefield)? Against terrorists? Against rioting prison inmates?

3. How do you think other nations would react to the first use of incapacitating agents?

Difficult Decisions

Reproductive Technologies

Tangling the roots of the family tree.

by Jal S. Parakh and Irwin L. Slesnick

Infertile couples, desperate to start a family, commonly turn to adoption. But because many couples want children with certain traits, the decision to adopt often means years of waiting and, ultimately, the frustration of life without children. The development of reproductive technologies, such as artificial insemination, has given these childless couples a new option. The first successful artificial insemination probably occurred late in the 18th century, when John Hunter, an English physician, inseminated a woman with semen from her husband. Nowadays, 2000 babies are born each year in the United States as a result of artificial insemination with sperm from a husband (AIH) or other donor (AID).

Until a decade ago, little could be done for women who could not ovulate or whose fallopian tubes had blockages that prevented sperm from reaching and fertilizing eggs. Then, in 1978, headlines around the world announced the historic birth of the world's first "test-tube baby." At 11:47 PM on July 25, John and Lesley Brown became the ecstatic parents of Louise Brown, born at Oldham General Hospital in England. The birth of baby Louise was the fruition of the pioneering research of gynecologist Patrick Steptoe and embryologist Robert Edwards in the technique of *in vitro* fertilization (IVF). In IVF, doctors surgically remove eggs from a woman's ovaries and place those eggs, along with a man's sperm, in a sterile glass dish, where fertilization takes place. After several days, doctors inject the embryos into a woman's uterus. The embryos implant in the uterine wall and an otherwise normal pregnancy begins.

In 1983, Elsa and Mario Rios died in a plane crash, and left the world with perplexing ethical and legal problems—two embryos in cold storage in Queen Victoria Medical Center in Melbourne, Australia. Before the couple's death, eggs had been removed from Elsa Rios and fertilized with sperm from an anonymous donor. While waiting to see if the first implant took, doctors had frozen the two extra embryos. Should the embryos be destroyed now that the couple is dead? Should the embryos be implanted in other childless women who want babies? The estate of Mario and Elsa Rios was estimated at about a million dollars. Who should inherit their million-dollar estate? Is one of the eight-cell embryos a millionaire?

The new reproductive technologies are making us face troubling questions that go to the very heart of long-held and cherished beliefs about the meaning of parenthood and about the family as a unit for our society. The resulting complicated ethical and legal problems are currently the subject of heated debates. From a utilitarian view, reproductive technology is not inherently objectionable as long as it is likely to lead to more good than harm. This view argues that the marital bond may even be strengthened by the birth of a child. Another argument for the use of artificial insemination and IVF is that every doctor has a duty to help infertile couples conceive, thus satisfying their strong, natural urge to have children.

Some people feel, however, that any intervention to help couples have children using the new reproductive technologies is wrong. They argue that these technologies not only destroy the exclusivity of the marital relationship, but also weaken the natural bond between parents and the child born as a result of AID or IVF.

Jal S. Parakh and Irwin L. Slesnick are professors in the biology department at Western Washington University, Bellingham, WA 98225.

Questions for discussion

You may find the questions on the accompanying student page (see page 21) sufficient to provoke discussion and analysis of the ethics of the new reproductive technologies. If further questioning is needed to stimulate deeper explorations of the ideas, you might ask such questions as the following:

- IVF may lead to a eugenics program that produces "superior" humans by combining donor sperm and donor eggs from people with highly desirable and useful traits, such as high intelligence and great strength. What would be the consequences of such a program for the "super children" and for our society?
- Some people argue that researchers could learn a great deal by using human embryos produced by IVF specifically for research purposes. Others argue that such research is dehumanizing and should not be permitted. Would you allow these embryos to be used in research?
- What do you think should be done with the leftover frozen embryos of Mario and Elsa Rios, or of any another couple who died before completing the IVF procedure?
- Suppose a husband wants eggs removed from his dying wife's body to be fertilized with his sperm *in vitro* and then implanted in a surrogate mother. Should we allow this request?
- Suppose a financially independent single woman wishes to become a mother but does not wish to conceive by intercourse and requests artificial insemination by an unknown donor at a sperm bank. Should we grant her request? Would it make a difference if she is a lesbian?
- In some cities, men and women can sell blood legally. Should men be permitted to sell sperm, and women eggs? ■

For further reading

Kramer, M. (1985) "Last Chance Babies: The Wonders of In Vitro Fertilization." *New York*, 18(31):34–42.

McCormick, R.A. (1985) "Therapy or Tampering? The Ethics of Reproductive Technology." *America*, 153(17):396–403.

Munson, R. (1988) *Intervention and Reflection: Basic Issues in Medical Ethics*. Bellmart, Calif.: Wadsworth, 1988.

Wallis, C. (1984) "The New Origins of Life." *Time*, 124(11):46–53.

Should We Redefine Parenthood?

—Art by Max-Karl Winkler

Maybe you know people who've announced their wedding engagements in the paper. Would you be surprised if you read something like the following under their picture?

> Adrien Mildred Rensberger and Allen Stephen Pratt announce their engagement. The bride-to-be is the daughter of Mary Kay Molinski, Sara Jean McIntyre, Samuel Terrence Kelly, Carol Lee Rensberger, and James S. Rensberger. The future bridegroom is the son of . . .

In vitro fertilization (IVF) has made it possible for one child to have five different parents. In this complex family, the donor mother would contribute the egg. A second, surrogate mother would carry the developing child in her uterus through 9 months of pregnancy. One father would contribute the sperm for fertilization. Finally, a third mother and a second father would raise the child as their own. Since we now can freeze sperm, eggs, and embryos for future use, IVF families may become even more complex.

While IVF solves some people's infertility problems, it also raises new ethical issues. What do you think might be the social consequences of changing the nature of the biological family?

Questions to consider

1. Imagine that you are a man whose wife is infertile. Her oviducts are blocked and eggs cannot move from the ovary to the uterus. You learn that a nearby clinic can remove an egg from an ovary, fertilize the egg in a dish, and then implant the embryo in your wife's uterus. Would you accept this procedure for reproduction even though it is unnatural? Do you think you would feel different if you were the wife?
2. Now suppose you are a woman who is fertile but cannot survive a pregnancy. Your husband is fertile. Your sister volunteers to carry an embryo conceived by IVF without making any legal claims on you. How do you feel about the ethics of this procedure?
3. Suppose that you are part of a couple in which the husband cannot produce living sperm. You and your spouse decide to buy frozen sperm. Your baby will be conceived by artificial insemination. Do you think this action is ethical?
4. Do you feel that IVF could go beyond boundaries healthy for society? What laws would you write to protect society against the abuses of IVF?

Difficult Decisions: Genetic Screening

Do the benefits outweigh the drawbacks?

by Irwin L. Slesnick and Jal S. Parakh

Many genetic disorders can be detected with tests of blood and chromosomes. *Genetic screening* is the large-scale use of these tests as part of public health programs. By 1909, physician Archibald Garrod had provided the biochemical knowledge that ultimately led to the screening of newborns.

Garrod's pioneering work helped to establish that alkaptouria and other metabolic diseases were due to the lack of certain enzymes necessary to catalyze the breakdown of amino acids, which then accumulated in the patients. Garrod also proposed that such disorders could be inherited as Mendelian recessive traits.

PKU, a similar "inborn error of metabolism" was clearly identified by Fölling in 1934. Diagnostic tests for PKU have been available since 1934, but early versions were time-consuming and costly. In 1962, Dr. Robert Guthrie developed a faster, cheaper, automated testing procedure for PKU. The possibility of identifying PKU children and treating them with a preventive diet led Massachusetts to pass the first *mandatory* PKU screening law in 1963. By 1967, 41 states had passed similiar laws.

Genetic screening reached a new stage in the late 1960s when public health officials initiated *voluntary* screening programs of adults to detect heterozygous carriers of specific recessive traits, such as Tay-Sachs disease. Subsequently, several states passed laws to establish *mandatory* programs to screen for carriers of sickle-cell trait.

The question of whether genetic screening should be voluntary or mandatory is a very controversial issue. So is the goal of the screening whether to detect *afflicted individuals* or *carriers*. Screening can be performed at various life stages, such as prenatal, neonatal, newborn, and premarital.

Different members of society have advocated genetic screening to achieve different goals. Preventive, or negative, eugenics aims at reducing the frequency of "undesirable" traits in human populations by sterilizing "unfit" individuals or aborting "unfit" fetuses. Genetic screening is more commonly used to provide individuals with the information and genetic counseling needed to make decisions about having children, aborting fetuses, and getting timely treatment for afflicted newborns.

Most states recognize their obligation to provide some genetic screening for newborns, especially for such defects as PKU and hypothyroidism, which respond well to therapy. These programs are easy to carry out and are acceptable to the public. But the expansion of genetic screening to cover all newborns and a larger pool of adults for a range of treatable genetic diseases raises a number of issues.

We need to calculate the economic costs and benefits of comprehensive screening of newborns. We must also consider the potential consequences—such as emotional ones—of genetic disease on the family and society; the potential loss of privacy to tested individuals; and the use or abuse of the information by schools, employers, insurance companies, and other private or governmental agencies.

Questions for discussion

The questions on the accompanying student page (see page 29) may be sufficient to provoke discussion about genetic screening. The following questions could stimulate deeper thinking on the issue:

- Genetic screening is not foolproof. Some tests can only show the probability that a person is a carrier. Does this weakness affect your thinking on genetic screening? How?
- Will mandatory genetic screening programs lead to laws that govern a person's reproductive behavior?
- Whose rights could be violated if genetic screening were mandatory?
- Should members of high-risk ethnic groups be screened selectively, for example, Ashkenazi Jews for Tay-Sachs, blacks for sickle-cell anemia, Scandinavians for PKU, Italians and Greeks for thalassemia?
- Some minority ethnic groups regard genetic screening, which seeks out individuals who should not reproduce, as a thinly veiled genocidal program of majority ethnic groups. What do you think? ■

For further reading

Beauchamp, T.L., and L. Walters. *Contemporary Issues in Bioethics.* Encino, Calif.: Dickenson, 1978.

Knowles, R.V. *Genetics, Society, and Decisions.* Columbus: Merrill, 1985.

Munson, R. *Intervention and Reflection: Basic Issues in Medical Ethics.* Belmont, Calif.: Wadsworth, 1988.

President's Commission for the Study of Ethical Problems in Medical and Biomedical and Behavioral Research. *Screening and Counseling for Genetic Conditions. A Report on the Ethical, Social, and Legal Implications of Genetic Screening, Counseling, and Education Programs.* Washington, D.C.. Government Printing Office, 1983.

Rowley, P.T. "Genetic Screening: Marvel or Menace?" *Science,* 225:138–144, July 13, 1984.

Stine, G.J. *Biosocial Genetics: Human Heredity and Social Issues.* New York: Macmillan, 1977.

Irwin L. Slesnick and Jal S. Parakh are professors in the biology department at Western Washington University, Bellingham, WA 98225.

Should We Genetically Screen Newborns?

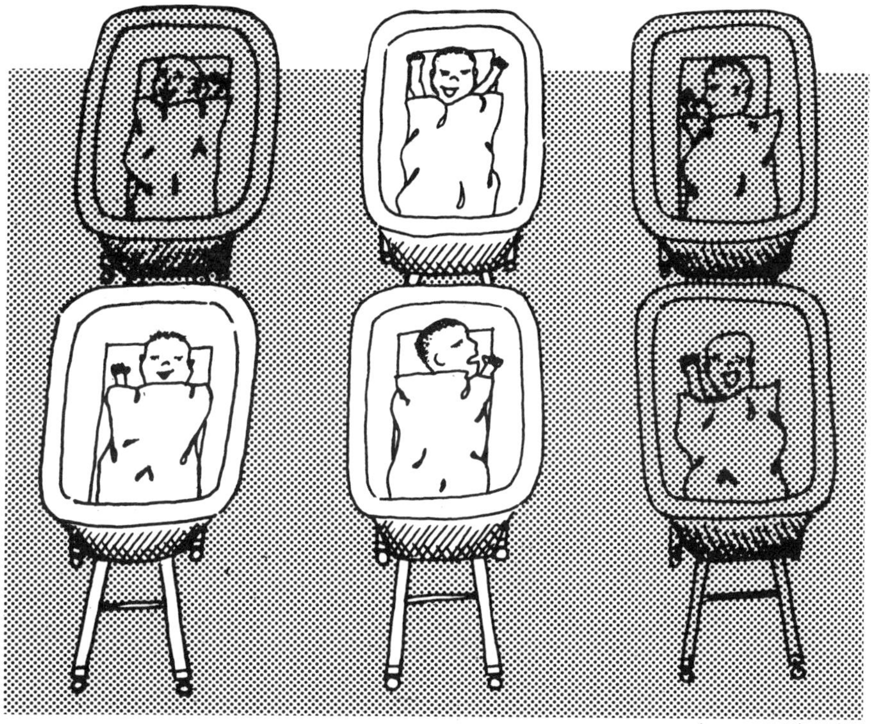

—Art courtesy of the authors

The genes on our chromosomes play a major role in making us the way we are. *Genetic screening* is the testing of people for particular genes, especially those genes that may cause diseases. More than 3000 human disorders involve genetic factors.

Screening can be carried out at any time in the life cycle. People might be screened before birth, immediately after birth, in school, before marriage, or even before employment. In a few years, scientists expect to be able to provide people with total gene screens, complete printouts of their approximately 100 000 genes.

The newborns in the maternity nursery shown above were genetically screened by blood and chromosome tests. Today's test found three potentially serious health problems.

Baby #1: This infant has an underactive thyroid gland (hypothyroidism). Replacement hormone treatment will prevent her from becoming seriously retarded.

Baby #2: This girl is heterozygous for sickle-cell anemia; that is, she is a carrier. If she later has a child with a man who is also a carrier of this genetic condition, there is a 25 percent chance that the child will suffer from the deadly disease.

Baby #3: This boy has a sex-linked chromosomal defect, known as fragile X, that can cause mental retardation and other symptoms. Doctors are treating fragile-X syndrome children successfully with experimental drugs.

Imagine one of the babies with health problems is yours. What would your reaction be to the results of the screening?

Questions to consider

1. Do you think genetic screening of newborns should be mandatory or voluntary?
2. Whether screening is mandatory or voluntary, who should have access to information about a person's genetic makeup? The tested individual? Parents? Doctors? Schools? Governments? Insurance companies? Employers?
3. Should the state require that children who test positive for hereditary diseases undergo treatment to correct genetic defects?
4. Suppose you were screened as an infant. The test showed that you had a dominant gene for a fatal disease with delayed onset, such as Huntington's disease. When would you want to know about your condition? Why do you feel this way about personal genetic information?

Difficult Decisions: AIDS

The first in a series of thinking activities.

by Irwin L. Slesnick

Whether the AIDS epidemic will be the Black Death of modern times is still uncertain. The World Health Organization predicts that 1 out of every 50 people could be infected with the AIDS virus by 1990. The course of the AIDS epidemic in the world population depends on how the disease will sustain itself through heterosexual transmission and whether people will change their sexual practices.

With the near certain knowledge of how the disease is transmitted, public health professionals are confident the epidemic can subside in both homosexual and heterosexual communities through forthright programs of sex education. The power of public health education in extending life and in raising its quality is historic.

Accompanying the spread of the AIDS virus is a second epidemic of fear, anxiety, and torment. According to several public opinion polls, more than 100 million Americans have an "acute" fear of AIDS. A CBS poll in September 1985 showed that 47 percent of Americans thought they could catch AIDS from a drinking glass. About 30 percent thought the virus could be transmitted by kissing or from a toilet seat. A Harris poll in late October 1985 disclosed that 50 percent of Americans thought that AIDS spreads by casual contact. And in January 1987, a *Wall Street Journal* poll reported that 24 percent of Americans thought children with AIDS should not attend school with other children.

A consequence of the fear and hysteria about AIDS is the addition of another level—an unnecessary level—of suffering to the lives of AIDS victims and carriers. People known or just suspected to be infected with the AIDS virus have been fired from jobs, evicted from apartments, discharged from the military, and rejected by friends and family. Even some physicians, hospitals, and ambulance crews deny them service. Yet all evidence thus far analyzed indicates that infection by casual contact has not occurred.

Like other responsible researchers, health scientists are reluctant to make absolute statements about AIDS. They do not use such comforting terms as "certainly," "never," and "no-risk." Rather, they will say "data indicate," "highly probable," and "apparent low risk." To people unfamiliar with the nature and language of science and who go through life with white knuckles, this caution fails to dispel their fearful fantasies.

Of the approximately 40 000 AIDS cases diagnosed in America so far, no one is known to have passed the virus to a family member. We should be able to gain some security in the knowledge that people living in the same household seem to be safe. Nevertheless, even remote risks of infection are reason enough for some people to insist on excluding AIDS victims from public life. They think it is better to err on the side of extreme caution. While segregating lepers and Japanese Americans may have made a hysterical public feel safer, history demonstrates that in exchange for false security, society committed some shameful acts of cruelty against innocent and harmless humans.

Questions for discussion

You may find the questions on the accompanying student page (see page 35) sufficient to provoke discussion and analysis of public fear of AIDS. If further questioning is needed to stimulate deeper explorations of the ideas, you might ask such questions as the following:

- How do newspapers and magazines reinforce public fear and hysteria with such headlines as "School Cook Dies of AIDS" and "Now No One Is Safe From AIDS"? (The first headline implies that school children who ate the cook's food are at risk. The second headline fails to add that no one *who has sex with a carrier or who receives contaminated blood* is safe.)
- How much do the people in your community know about AIDS? How do you think they feel about the civil rights of people with AIDS? Test your ideas about public knowledge and public opinion about AIDS with a questionnaire of about five or six questions.
- Should everyone be screened for AIDS antibodies to determine whether they have been infected by the AIDS virus?
- Suppose as director of the World Health Organization you were given unlimited funds to stop the AIDS epidemic. Into what project(s) would you put the money?
- The AIDS virus probably originated several decades ago as a mutation in a harmless virus in the green monkey of central Africa. Mutations of viruses in the cells of birds and infrahuman mammals are thought to be responsible for many varieties of human influenza virus. How can humans protect themselves from epidemics caused by mutant microorganisms? (Mutant organisms are often the cause of epidemics, such as AIDS or swine flu. In many individuals, the immune system cannot respond to new infectious organisms. To avoid devastating epidemics, society must be responsive to measures of prevention—for example, changes in sexual practices—and cure—for example, isolation of mutant organisms and the production of vaccines.)
- In what ways is the AIDS epidemic similar to and different from the Black Death epidemic, which killed 25 percent of the population of Europe in the 14th century? ■

Irwin L. Slesnick is a professor in the biology department at Western Washington University, Bellingham, WA 98225.

For further reading

"AFRAIDS." *The New Republic,* 193:7–9, October 14, 1985.

Bastian, B.J. "AIDS: Education Against Fear." *Carolina Tips,* vol. 49, no. 2.

Kolata, G. "Mathematical Model Predicts AIDS Spread." *Science,* 228:1464–1465, March 20, 1987.

Reed, S. "AIDS in the Schools: A Special Report." *Phi Delta Kappan,* 67:494–498, March 1986.

Is Jim Still Your Friend?

—*Art by Mignon Khargie*

Jim has AIDS. The AIDS virus is destroying his white blood cells. Without these blood cells, he cannot fight off diseases caused by germs or cancer cells. While new drugs may prolong his life, he will likely die of a disease his weakened immune system cannot overcome.

The AIDS virus is reproducing inside his white blood cells. His circulatory system carries the virus to all the fluids in his body. The AIDS viruses are in Jim's blood, tears, and semen; to a far lesser concentration, they are also in his saliva, his urine, and the fluids between the cells of his body organs.

Studies of the spread of AIDS show that in order for an AIDS victim to infect others, some of her or his infected body fluid must enter the bloodstream of the other person. This transfer of infected fluid can happen during sexual acts, blood transfusions, or the sharing of hypodermic needles. Infection can occur through contact between open sores or by biting. AIDS is not spread by casual contact. The normal, everyday interactions between people cannot spread the virus.

However, public opinion polls show that more than 50 percent of Americans fear that AIDS can be spread by casual contact—drinking from the same glass, kissing, using the same toilet seat, or even working in the same room. Some people think AIDS victims and carriers should be isolated from the uninfected portion of society.

Questions to consider

1. You would normally invite Jim to the wedding reception your family is planning. Will you invite him? Whether your answer is yes or no, how will you explain your decision to yourself? To your family? To your friends? Explain why you feel as you do.

2. Often people have hysterical reactions to situations that appear life-threatening. Sometimes ignorance increases the fear. Should the public be given all the information available even if the information is frightening?

Difficult Decisions: Human Cryonics

by John A. Miller and Irwin L. Slesnick

A hot debate on a cold topic.

In the 17th century, chemist Robert Boyle made the first scientific study of the effects of cold on organic matter, using fruit and eggs as his test subjects. The first experiment in which animal cells were frozen and then proved to be viable upon thawing did not occur until 1949. That year, three British cryobiologists found that fowl spermatozoa remain viable if placed in antifreeze, or *cryoprotectant*, prior to being frozen in a coolant. The researchers used a solution of 1.4 M salt-buffered glycerol as their cryoprotectant and solid carbon dioxide at –76°C as their coolant. Glycerol infiltrates cells, thereby preventing formation of lethal ice crystals.

Today the procedure is common; for example, we freeze bull spermatozoa in liquid nitrogen at –196°C as a matter of convenience and economy for cattle breeding. A bank of frozen spermatozoa of Nobel prize winners is available to women desiring artificial insemination. Infertile couples undergoing embryo implantation can freeze eight-cell fertilized embryos in liquid nitrogen for later thawing and transfer into the womb if the first implantation does not succeed.

In a 1949 study, cryobiologist B. J. Luyet successfully preserved red blood cells with a cryoprotectant and liquid nitrogen. These blood cells are useful for blood transfusions and serological testing. Further studies in the early 1950s by D.W.H. Barnes and J. F. Loutit demonstrated that bone marrow cells can survive freezing at –79°C when they are bathed in glycerol or dimethyl sulfoxide (DMSO). This technique is now used to provide bone marrow cells for the treatment of radiation damage.

Lymphocytes and platelets can also be frozen in liquid nitrogen using glycerol or DMSO as the cryoprotectant. Unfortunately, the survival rate for platelets after only 24 hours at liquid nitrogen temperatures (–196°C) is only 70 percent.

Skin subjected to cryonic conditions can be revitalized for use as skin grafts in burn victims, although there is always a risk of infection due to freezing-related death of some of the cells. Corneas can be preserved in liquid nitrogen for later transplants. Bone and its marrow (but not attached cartilage) are also being frozen successfully for later use.

Less success has been obtained with other whole tissues. Hamster hearts and rat hearts stored at –20°C for 20 minutes show reasonable survival, but none survive when stored at –79°C. Dog liver stored 2 weeks at –60°C has poor transplant success.

Failure at freezing major organs such as hearts and livers would seem to preclude any success with whole animals. The evidence is mixed on this point, however. English scientist Audrey Smith had great success in the early 1950s freezing hamsters. She lowered the hamsters' body temperature to –5.5°C for up to 4 hours with a 100 percent recovery rate. In contrast, rabbits subjected to similar conditions exhibited very low survival rates. And, at lower temperatures, even the hamster recovery rate dropped. When temperatures are reached in which 55 percent of the hamsters' body water froze, no hamsters survived.

The studies involving the freezing of dogs are the most intriguing. In 1986, Paul Segall and Hal Sternberg at the University of California at Berkeley replaced the blood in a dog named Miles with a cryoprotectant, as they lowered the dog's temperature to 5°C. They held him at this temperature for 20 minutes and then reversed the process. Miles survived and today seems to be a normal, healthy dog.

The Alcor Life Extension Foundation in California has performed related experiments with 12 dogs. These dogs were kept at lowered temperatures for up to 4 hours. Immediately after thawing, 6 of the dogs died or developed seizures. Only 6 dogs actually recovered for any period of time, and only 1 of these seemed to be normal a year after being thawed.

Cryonics technology is already being applied to humans. At present, there are at least 5 people known to be frozen in liquid nitrogen, and the number may be as high as 12 or more. Why would anyone wish to be preserved in this manner? The answer lies in the search for a chance at immortality. Such people hope that they can be revived at a time when aging processes are under scientific control, or, if they were ill when frozen, when there is a cure for the diseases that were killing them.

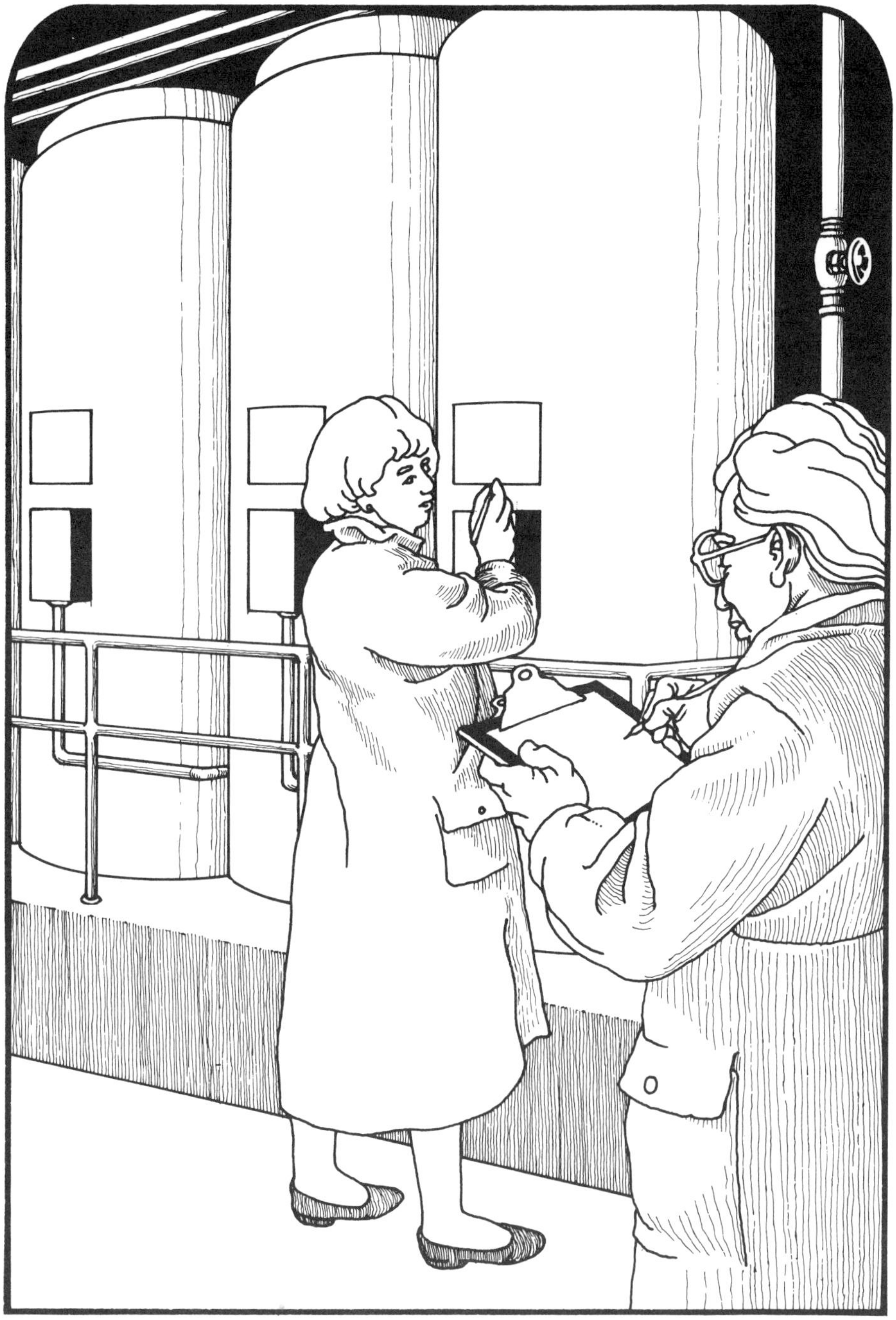

—*Art by Max-Karl Winkler*

Questions for discussion

You may find the questions on the accompanying student page (see page 35) sufficient to provoke discussion and analysis of the ethics of human cryonics. If further questioning is needed to stimulate deeper explorations of the ideas, you might ask such questions as the following:

- Is freeze preservation of living humans an acceptable practice for any of these groups of people: astronauts on space flights requiring many years of travel time; terminally ill people who hope the future will bring a cure for their illnesses; depressed people who choose freezing instead of suicide; curious people who want to freeze, thaw, and see what the future's like; or dangerous criminals for whom there are now no proven methods of rehabilitation?
- What might be some of the problems facing a society that has the technology to successfully freeze and later revive humans? What might be some of the advantages of human cryonics?
- Do people have the right to achieve "immortality" and thus add to the global population later? ■

John A. Miller is a professor in the chemistry department and Irwin L. Slesnick is a professor in the biology department at Western Washington State University, Bellingham, WA 98225.

For further reading

Fennema, O.R., W.D. Pamrie, and E.H. Marth. *Low-Temperature Preservation of Foods and Living Matter.* New York: Marcel Dekker, 1973.

Freifeld, K. "Fresh-frozen." *Forbes,* 135:90–91, January 28, 1985.

Kahn, C. "On Ice." *Health,* 19:70–73, March 1987.

Smith, A.A., ed. *Current Trends in Cryobiology.* New York: Plenum Press, 1970.

"Today a Frozen Dog, Tomorrow the Iceman." *Discover,* 8:9, June 1987.

Vogel, S. "Cold Storage." *Discover,* 9:52–54, February 1988.

Weisburd, S. "Beyond the Cutting Edge of Cold." *Science News,* 132:138–141, August 29, 1987.

If We Freeze Humans Today, What Will Happen To Society Tomorrow?

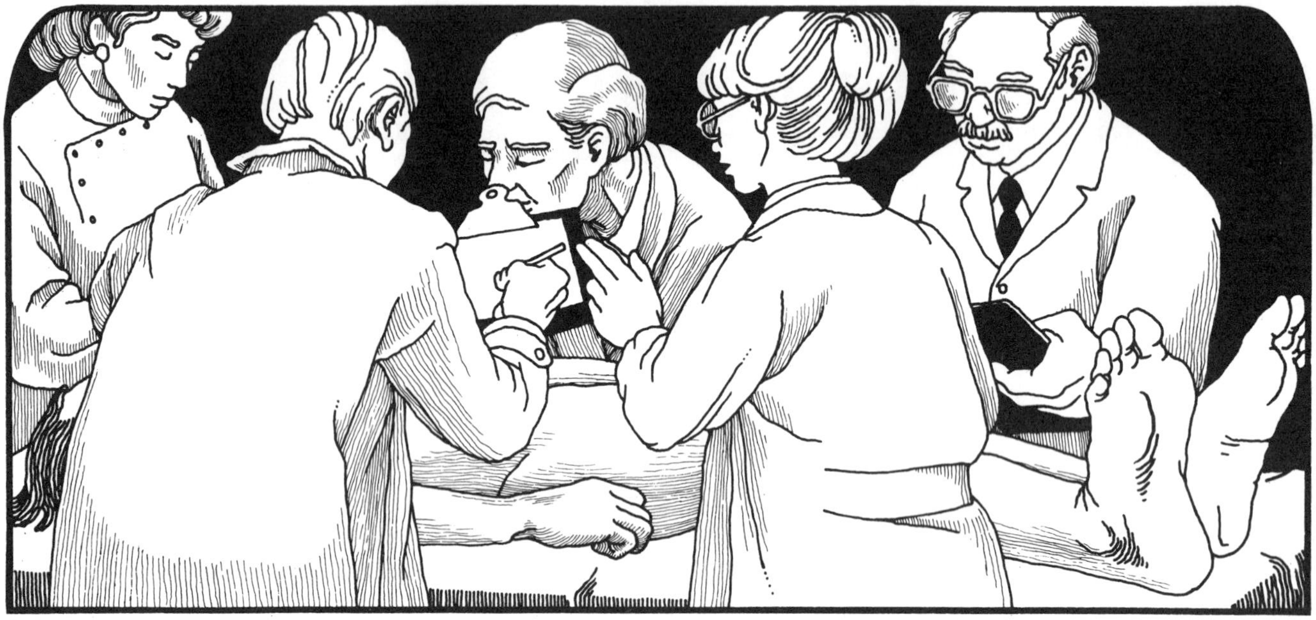

—*Art by Max-Karl Winkler*

In recent years, experiments have been performed in which living animals are frozen and then successfully revived. In 1986, scientists replaced the blood in a dog named Miles with a type of antifreeze solution and lowered his body temperature to 5°C. The antifreeze solution prevented lethal ice crystals from forming in the cells of his body. After holding Miles at this temperature for 20 minutes, the scientists reversed the process and revived him. Miles is still alive and healthy.

The science of freezing and reviving animals is called *cryonics*. Given the success of the Miles experiment, you might expect to read something like the following in the newspapers of the near future:

> *March 2, 2000. New York, NY.*— Today marks a revolutionary step in the treatment of disease—the first use of cryonics by a major hospital. Doctors at Central Hospital declared Laura Kay, 44, victim of an incurable heart condition, incurably and terminally ill today at 1:32 PM. By 1:33 PM, they had replaced most of her blood with a kind of antifreeze and begun cooling her body to subzero temperatures.
>
> After freezing, Laura Kay's heart stopped beating and her brain no longer produced brain waves. She was therefore legally dead—but actually, she was in a state of suspended animation. Laura Kay has arranged to remain frozen until doctors discover a cure for her fatal disease; she will then be thawed and have her condition corrected.

How will society respond to the problems posed by the freeze preservation of whole human bodies?

Questions to consider

1. How should freezing followed by legal death and later reanimation affect the status of a person's spouse, children, estate, and debts?
2. Do you think Laura Kay should legally be considered a new citizen after she is thawed?
3. Who will love Laura Kay when she reappears? Who will take care of her? How will she get along in a future society? If it takes 100 years for a treatment for Laura Kay's heart defect to be found, should her great, great, great grandchildren have to be responsible for her after she thaws?